Energy Sector Standard of the People's Republic of China

NB/T 31011-2019

Replace NB/T 31011-2011

Preparation regulation for cost estimation of onshore wind power projects

陆上风电场工程设计概算编制规定及费用标准

(English Translation)

China Water & Power Press

中国水利水电出版社

Beijing 2024

图书在版编目（CIP）数据

陆上风电场工程设计概算编制规定及费用标准 : NB/T 31011-2019代替NB/T 31011-2011 = Preparation regulation for cost estimation of onshore wind power projects（NB/T 31011-2019 Replace NB/T 31011-2011） : 英文 / 国家能源局发布. -- 北京 : 中国水利水电出版社, 2024. 8. -- ISBN 978-7-5226-2690-1

Ⅰ. TM614

中国国家版本馆CIP数据核字第20240QT615号

Energy Sector Standard of the People's Republic of China

中华人民共和国能源行业标准

Preparation regulation for cost estimation

of onshore wind power projects

陆上风电场工程设计概算编制规定及费用标准

NB/T 31011-2019

Replace NB/T 31011-2011

（English Translation）

Issued by National Energy Administration of the People's Republic of China

国家能源局　发布

Translation organized by China Renewable Energy Engineering Institute

水电水利规划设计总院　组织翻译

Published by China Water & Power Press

中国水利水电出版社　出版发行

Tel: (+ 86 10) 68545888　68545874

sales@mwr.gov.cn

Account name: China Water & Power Press

Address: No.1, Yuyuantan Nanlu, Haidian District, Beijing 100038, China

http: //www.waterpub.com.cn

中国水利水电出版社微机排版中心　排版

北京中献拓方科技发展有限公司　印刷

210mm×297mm　16 开本　4 印张　161 千字

2024 年 8 月第 1 版　2024 年 8 月第 1 次印刷

Price（定价）: ¥ 640.00

About English Translation

This English version is one of China's energy sector standard series in English. Its translation was organized by China Renewable Energy Engineering Institute authorized by National Energy Administration of the People's Republic of China in compliance with relevant procedures and stipulations. This English version was issued by National Energy Administration of the People's Republic of China in Announcement [2023] No. 1 dated February 6, 2023.

This version was translated from the Chinese Standard NB/T 31011-2019, *Preparation regulation for cost estimation of onshore wind power projects*, published by China Water & Power Press. The copyright is reserved by National Energy Administration of the People's Republic of China. In the event of any discrepancy in the implementation, the Chinese version shall prevail.

Many thanks go to the staff from the relevant standard development organizations and those who have provided generous assistance in the translation and review process.

For further improvement of the English version, any comments and suggestions are welcome and should be addressed to:

China Renewable Energy Engineering Institute
No. 2 Beixiaojie, Liupukang, Xicheng District, Beijing 100120, China
Website: www.creei.cn

Translating organization:

China Renewable Energy Engineering Institute (China Renewable Energy Engineering Cost Management Center)

POWERCHINA Northwest Engineering Corporation Limited

Translating staff:

WANG Yonggang	LI Hongyu	ZHANG Xin	XUE Huijuan
WU Xiao	GUO Zhenni	ZHAO Guizhi	YAO Yajiao
LI Kejia	LIU Qing	YANG Yuanzhao	

Review panel members:

QIE Chunsheng	Senior English Translator
LI Dongwei	POWERCHINA Huadong Engineering Corporation Limited
LI Zhongjie	POWERCHINA Northwest Engineering Corporation Limited
GUO Jie	POWERCHINA Beijing Engineering Corporation Limited
JIA Haibo	POWERCHINA Kunming Engineering Corporation Limited
LIANG Hongli	Shanghai Investigation, Design & Research Institute Corporation Limited
LI Yu	POWERCHINA Huadong Engineering Corporation Limited
ZHU Mingrun	POWERCHINA Chengdu Engineering Corporation Limited
ZHANG Qing	POWERCHINA Zhongnan Engineering Corporation Limited
LI Shisheng	China Renewable Energy Engineering Institute

National Energy Administration of the People's Republic of China

翻译出版说明

本译本为国家能源局委托水电水利规划设计总院按照有关程序和规定，统一组织翻译的能源行业标准英文版系列译本之一。2023 年 2 月 6 日，国家能源局以 2023 年第 1 号公告予以公布。

本译本是根据中国水利水电出版社出版的《陆上风电场工程设计概算编制规定及费用标准》NB/T 31011—2019 翻译的，著作权归国家能源局所有。在使用过程中，如出现异议，以中文版为准。

本译本在翻译和审核过程中，本标准编制单位及编制组有关成员给予了积极协助。

为不断提高本译本的质量，欢迎使用者提出意见和建议，并反馈给水电水利规划设计总院。

地址：北京市西城区六铺炕北小街 2 号
邮编：100120
网址：www.creei.cn

本译本翻译单位：水电水利规划设计总院（可再生能源定额站）
中国电建集团西北勘测设计研究院有限公司

本译本翻译人员：王永刚　李宏宇　张　鑫　薛惠娟
吴　霄　郭珍妮　赵桂芝　姚亚礁
李可佳　柳　青　杨元钊

本译本审核人员：

郄春生　英语高级翻译
李东伟　中国电建集团华东勘测设计研究院有限公司
李仲杰　中国电建集团西北勘测设计研究院有限公司
郭　洁　中国电建集团北京勘测设计研究院有限公司
贾海波　中国电建集团昆明勘测设计研究院有限公司
梁洪丽　上海勘测设计研究院有限公司
李　瑜　中国电建集团华东勘测设计研究院有限公司
朱明润　中国电建集团成都勘测设计研究院有限公司
张　庆　中国电建集团中南勘测设计研究院有限公司
李仕胜　水电水利规划设计总院

国家能源局

Contents

Foreword

This standard is drafted in accordance with the rules given in the GB/T 1.1-2009 *Directives for standardization—Part 1: Structure and drafting of standards.*

This standard replaces NB/T 31011-2011 *Specification and calculation basis for cost estimate of onshore wind power projects*. In addition to a number of editorial changes, this standard includes the following technical changes with respect to the NB/T 31011-2011.

—The scope of application of this standard is revised to "the preparation of cost estimation for new centralized onshore wind power projects, and this standard may be referenced in the preparation of cost estimation for other onshore wind power projects." (see 1 of this standard);

—In work breakdown, "preparation works for crane hardstand for wind turbine installation" (see 4.2.1.3 of this standard) and "safe and civilized construction measures" (see 4.2.1.5 of this standard) are, as first-level items, added to "auxiliary works". "Water supply for construction" (see 2.2.1.3 of NB/T 31011-2011) is deleted from first-level items and put under "other auxiliary works" (see 4.2.1.4 of this standard);

—In work breakdown, "collection line equipment and installation" (see 4.2.2.2 of this standard) is, as first-level item, added to "equipment and installation works". "Wind turbine outgoing lines" (see 4.2.2.1 of this standard), "subsystem testing" (see 4.2.2.3 of this standard), "electric system commissioning" (see 4.2.2.3 of this standard), "special electrical test" (see 4.2.2.3 of this standard), and "equipment allocation in control center" (see 4.2.2.4 of this standard) are added as second-level items. First-level item "control and protection equipment and installation" (see 2.2.2.3 of NB/T 31011-2011) as well as second-level items "safety monitoring equipment" (see 2.2.2.4 of NB/T 31011-2011), "environment and water and soil conservation equipment" (see 2.2.2.4 of NB/T 31011-2011), "national wind power information reporting system" (see 2.2.2.4 of NB/T 31011-2011), and "wind farm operation management system" (see 2.2.2.4 of NB/T 31011-2011) are deleted;

—In work breakdown, "collection line works" (see 4.2.3.2 of this standard) is, as first-level item, added to "civil works". "Wind turbine outgoing line project" (see 4.2.3.1 of this standard), "foundation works of reactive power compensators" (see 4.2.3.3 of this specification) and "flood (tide) control works" (see 4.2.3.5 of this standard) are added as secondary items. "Access roads and bridges" are merged into "access roads" in "transportation" (see 4.2.3.4 of this standard). First-level item "building" (see 2.2.3.3 of NB/T 31011-2011) is deleted, and the subordinate items are put under "step-up substation" (see 4.2.3.3 of standard). Second-level item "others" (see 2.2.3.5 of NB/T 31011-2011) is deleted;

—In work breakdown, "preliminary work cost" (see 4.2.4.2 of this standard) is, as first-level item, added to "other costs". "Project quality inspection and testing fee" (see 4.2.4.3 of this standard), "norm preparation and management cost" (see 4.2.4.3 of this standard), and "as-built drawing preparation cost" (see 4.2.4.5 of this standard) are added as second-level items. Second-level item "preliminary work cost" (see 2.2.5.2 of NB/T 31011-2011) is deleted. "Cost of management devices" and "purchase cost of tools, appliances and furniture for production" in "production preparation cost" are merged into "purchase cost of management tools, appliances and furniture" (see 4.2.4.4 of this standard);

—In cost composition, "staff welfare" and "labor protection expense" under "labor cost" are put under "enterprise accrual expense" (see 5.3.4.2 of this standard) in "indirect cost". "Measure cost" is modified into "other direct costs" (see 5.3.3.2 of this standard). The subordinate item "safe and civilized construction measures" is listed separately as second-level item under "auxiliary works" (see 4.2.1.5 of this standard). "Site management expense" and "overhead" in indirect cost are merged into "overhead" (see 5.3.4.1 of this standard). "Social security contributions" of production staff is included into labor cost (see 5.3.3.1 of this standard). "Norm preparation and measuring cost" (see 5.3.4.5 of this standard) is added. "Business tax" in taxes is changed into "value added tax" (VAT) (see 5.3.6 of this standard); "utilization cost of tools and equipment, insurance, taxes and additional education tax" is added into "overhead" (see 5.3.4.1 of this standard);

—In standard rates, "scientific research and test cost" (see 7.4.10 of this standard) and "investigation and design cost" (see 7.4.11 of this standard) are added. The skill levels of labor are changed from "highly-skilled labor, skilled labor, semi-skilled labor and unskilled labor" into "highly-skilled labor, skilled labor and unskilled labor", and the unit price of labor is adjusted from man-hour to man-day (see 7.1.1 of this standard). Additional construction cost for high-altitude areas (see 3.2.3.1 of NB/T 31011-2011) is deleted. Other related standard rates are adjusted according to the project categories, cost composition, calculation base of rate, market price level and the policy of replacing business tax with VAT.

National Energy Administration of the People's Republic of China is in charge of the administration of this standard. China Renewable Energy Engineering Institute has proposed this standard and is responsible for its routine management. Sub-committee on Planning and Design of Wind Power Project of Energy Sector Standardization Technical Committee on Wind Power is responsible for the explanation of the specific technical contents. Comments and suggestions in the implementation of this standard should be addressed to:

China Renewable Energy Engineering Institute
No. 2 Beixiaojie, Liupukang, Xicheng District, Beijing 100120, China

Drafting organizations:

China Renewable Energy Engineering Institute (China Renewable Energy Engineering Cost Management Center)

POWERCHINA Northwest Engineering Corporation Limited

Chief drafting staff:

GUAN Zongyin ZHAO Guizhi LI Hongyu HUANG Lin

XUE Huijuan WANG Yonggang SUN Lihong ZHANG Juan

This standard replaces NB/T 31011-2011.

Introduction

In recent years, with the rapid development of onshore wind power and the significant improvement in technology and equipment, China has introduced new regulations to increase labor income, improve working conditions, and enhance safety in production. Some new features and new situations have emerged in the construction process of wind power projects.

In order to promote the healthy and orderly development of onshore wind power industry, meet the requirements of project construction management, better reflect the relevant national policies, further strengthen and standardize the project cost management, improve the project cost management system, unify the preparation regulation and standard rates for cost estimation of onshore wind power projects, reasonably determine the project cost, and improve the investment returns, this standard has been developed according to the requirements of Document GNKJ [2015] No. 283 "Notice on Releasing the Development and Revision Plan of Energy Sector Standard in 2015" issued by National Energy Administration of the People's Republic of China, and after extensive investigation and research, summarization of practical experience, consultation of relevant standards of China, and wide solicitation of opinions.

Preparation regulation for cost estimation of onshore wind power projects

1 Scope

This standard specifies the work breakdown, cost composition, preparation of cost estimation, standard rates, and documentation of cost estimation for onshore wind power projects.

This standard is applicable to the preparation of cost estimation for new centralized onshore wind power projects, and may be referenced for other wind power projects.

2 Normative references

The following referenced document is indispensable for the application of this document. For dated references, only the edition cited applies. For undated references, the latest edition of the referenced document (including any amendments) applies.

NB/T 31010-2019, *Quota for cost estimation of onshore wind power projects*

3 General provisions

3.1 This standard is developed to unify the content, depth and calculation methods of cost estimation for onshore wind power projects, reasonably determine the costs, and improve the preparation quality of cost estimation.

3.2 This standard shall be used in conjunction with NB/T 31010-2019.

4 Work breakdown

4.1 General requirements

The work items for cost estimation of onshore wind power projects shall be broken down into auxiliary works, equipment and installation works, civil works and other cost items, as shown in Figure 1.

4.2 Work items

4.2.1 The auxiliary works refer to the temporary works and measures for the construction of main works, including temporary road works, power supply works, preparation works for crane hardstand for wind turbine installation, other auxiliary works, and safe and civilized construction measures. If there are auxiliary works for permanent use in civil works and equipment and installation works, they shall be listed under the corresponding items of permanent works.

4.2.1.1 The temporary road works refer to the temporary roads for the construction of wind power projects, including the construction, reconstruction (expansion) and rehabilitation of roads and bridges (culverts).

4.2.1.2 The power supply works refer to the connection of power supply lines of 10 kV and above from the existing power grid to the project site, and the preparation of power supply facilities of 35 kV and above.

4.2.1.3 The preparation works for crane hardstand for wind turbine installation refer to the site preparation works for the on-site assembly and installation of towers, wind turbines and other equipment.

4.2.1.4 Other auxiliary works refer to the works apart from the above, such as mobilization and demobilization of large hoisting machinery, water supply for construction, cofferdam construction, site grading for temporary facilities of wind farm in mountainous area. The mobilization and demobilization of large hoisting machinery refer to those Class A machinery included in the norms of machine-shift rates of construction machinery.

4.2.1.5 The measures for safe and civilized construction refer to the measures to be taken at the construction site by the contractor in accordance with corresponding requirements.

4.2.2 The equipment and installation works refers to all the equipment constituting the fixed assets of the wind farm and their installation, including power generation equipment and installation, collection line equipment and installation, step-up substation equipment and installation, and other equipment and their installation.

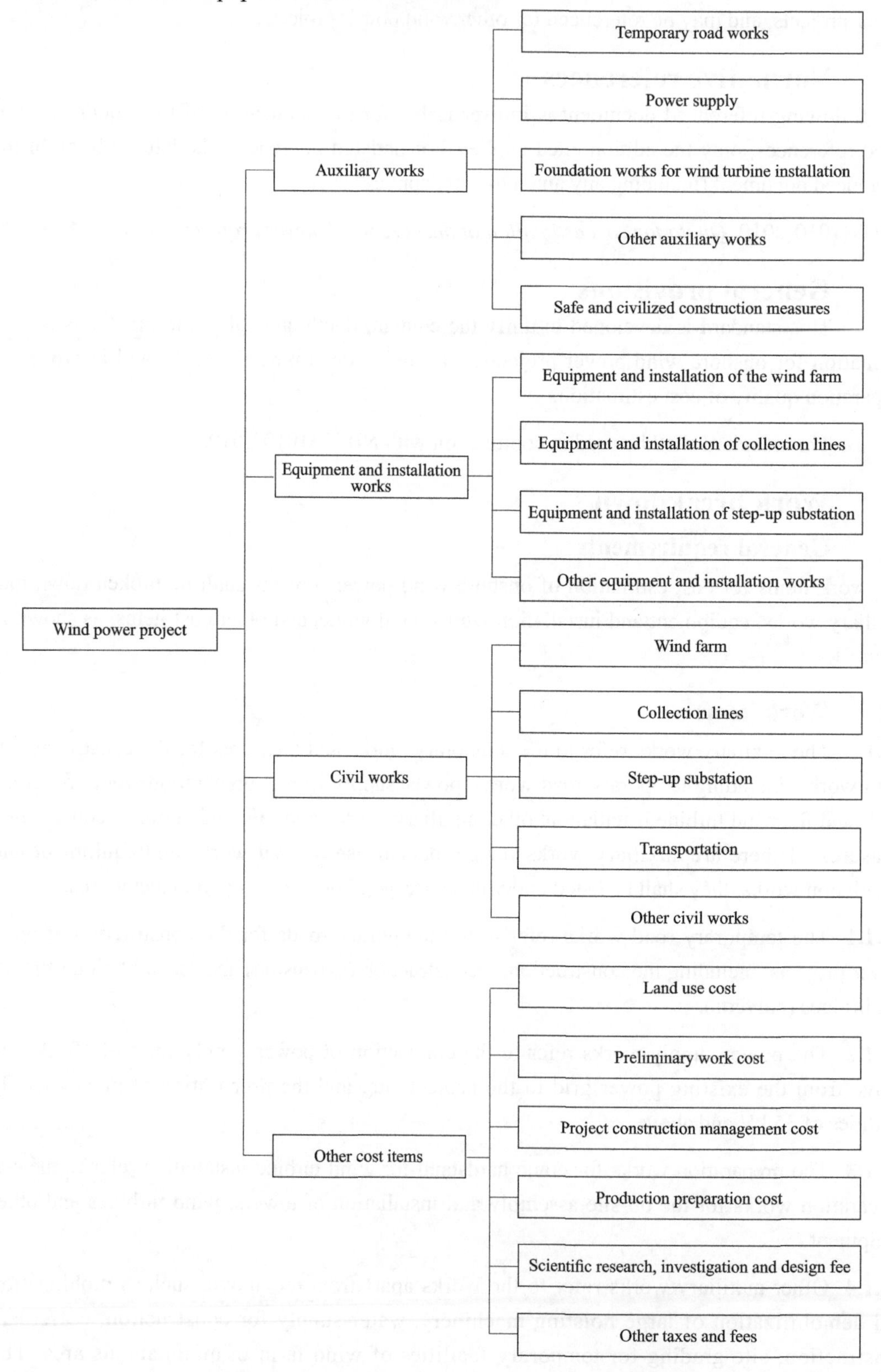

Figure 1 Work breakdown for cost estimation

4.2.2.1 The power generation equipment and installation refers to the wind turbines, towers, outgoing lines, transformers, and earthing and their installation.

4.2.2.2 The collection line equipment and installation refers to the collection cables, overhead collection lines, earthing, etc. and their installation.

4.2.2.3 The step-up substation equipment and installation refers to the equipment for power transformation, power distribution, control and protection in the step-up substation, including the main transformer system, power distribution equipment, reactive power compensation system, station service (emergency power supply) system, power cables, earthing, monitoring system, AC/DC system, communication system, remote control and metering system, subsystem testing, electrical system commissioning, and special electrical test.

4.2.2.4 Other equipment and installation refers to the corresponding equipment and installation other than mentioned above, including the heating, ventilation and air conditioning system, outdoor lighting system, firefighting system, water supply and drainage system, occupational health and safety equipment, production and maintenance vehicles, equipment allocation in control center, and other equipment that need be listed separately.

4.2.3 The civil works refers to the building (structure) works constituting the fixed assets of the wind farm, including wind farm works, collection line works, step-up substation works, road works, and other civil works.

4.2.3.1 The wind farm works refers to various building (structure) works in the wind farm, including the foundation works of wind turbines, outgoing line works, foundation works of wind turbine transformers, earthing of wind turbines and wind turbine transformers.

4.2.3.2 The collection line works comprises the civil works of the collection cable lines, the civil works of overhead collection lines, and overhead collection line earthing.

4.2.3.3 The step-up substation works refers to the structure works in the step-up substation, including the site grading, foundation works of main transformer, foundation works of reactive power compensators, foundation works of power distribution equipment, structures of power distribution equipment, production buildings, auxiliary buildings, site offices and living quarters, and outdoor works. Among the works, the production buildings comprise the central control room (building), distribution room (building), reactive power compensator room, etc. The auxiliary buildings comprise the sewage treatment room, fire pump room, firefighting equipment room, diesel generator room, boiler room, storehouse, garage, etc. The site offices and living quarters comprise the offices, duty room, dormitory, canteen, guard room, etc. The outdoor works comprise the walls, gates, roads, ground hardening, plantation, and other outdoor works which include the water supply pipes, drainage pipes, inspection wells, rainwater wells, sewage wells, well covers, valves, septic tank, and drainage ditches.

4.2.3.4 The road works refers to the access roads and site roads of the wind farm. The access roads refer to roads to the wind farm and the step-up station. The site roads refer to the roads for maintenance inside the wind farm.

4.2.3.5 Other civil works refers to those other than mentioned above, including the environmental protection, soil and water conservation, occupational health and safety, safety monitoring, water supply for firefighting, production and living, flood (tide) control, allocation of centralized production and operation management facility, and other works that need be listed separately.

4.2.4 Other cost items refer to the costs which are indispensable for the project but are excluded in equipment cost, installation cost, or civil works cost. Other cost items include the land use cost, preliminary work cost, the project construction management cost, production preparation cost, scientific research, investigation and design fee, etc.

4.2.4.1 The land use cost refers to the expenses paid in accordance with relevant national and local laws and regulations to obtain the site needed for project construction, including the land requisition cost, temporary land requisition compensation for attachments on the land, and site clearing expense.

4.2.4.2 The preliminary work cost refers to the expenses incurred in carrying out various works before the completion of the review of the pre-feasibility study report (or before the preparation of the wind farm project).

4.2.4.3 The project construction management cost refers to various administration expenses incurred in the process of project preparation, construction, commissioning, acceptance and handover, including the construction management cost, construction supervision cost, consulting service fee, techno-economic review fee, project quality inspection and testing fee, norm preparation and management cost, project acceptance cost, and insurance.

4.2.4.4 Production preparation cost refers to the expenses incurred for making neccssary preparation by the project legal person for project operation, including the cost of production staff training and advance mobilization into the plant, purchase cost of management tools, appliances and furniture, cost of spare parts, and cost of commissioning.

4.2.4.5 The scientific research, investigation and design fee refers to the expenses incurred in scientific research and test, investigation and design for the construction of the project, including the scientific research and test fee, investigation and design fee, and as-built drawing preparation cost.

4.2.4.6 Other taxes and fees refer to the taxes and fees to be paid in accordance with relevant national regulations, like the compensation for soil and water conservation, etc.

4.2.5 The work breakdown shall comply with Annex A of this standard. The level 2 and level 3 items in Table A.1 to Table A.4 may be added or deleted depending on the project design.

5 Cost composition

5.1 General requirements

The composition of total cost of an onshore wind farm is shown in Figure 2.

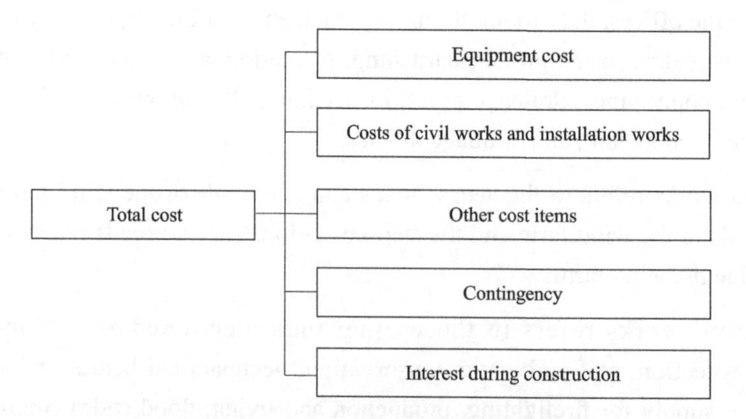

Figure 2 Composition of total cost

5.2 Equipment cost

5.2.1 The equipment cost consists of original price of equipment, freight and miscellaneous charges, transportation insurance, and purchase and care expense.

5.2.2 The original price of equipment:

a) The original price of domestic equipment refers to the factory price;

b) The original price of imported equipment consists of the cost, insurance and freight (CIF), tariff, VAT, agent charges, goods inspection fees and port charges.

5.2.3 The freight and miscellaneous charges refer to all expenses incurred in equipment transportation from factory to the site, including the expenses for transportation, dispatching, handling, packing and lashing, and other miscellaneous charges.

5.2.4 The transportation insurance refers to the premium paid for the transportation of equipment.

5.2.5 The purchase and care expense refers to various expenses incurred in the process of material purchase and storage.

5.3 Cost composition of civil works and installation works

5.3.1 The cost composition of civil works and installation works is shown in Figure 3.

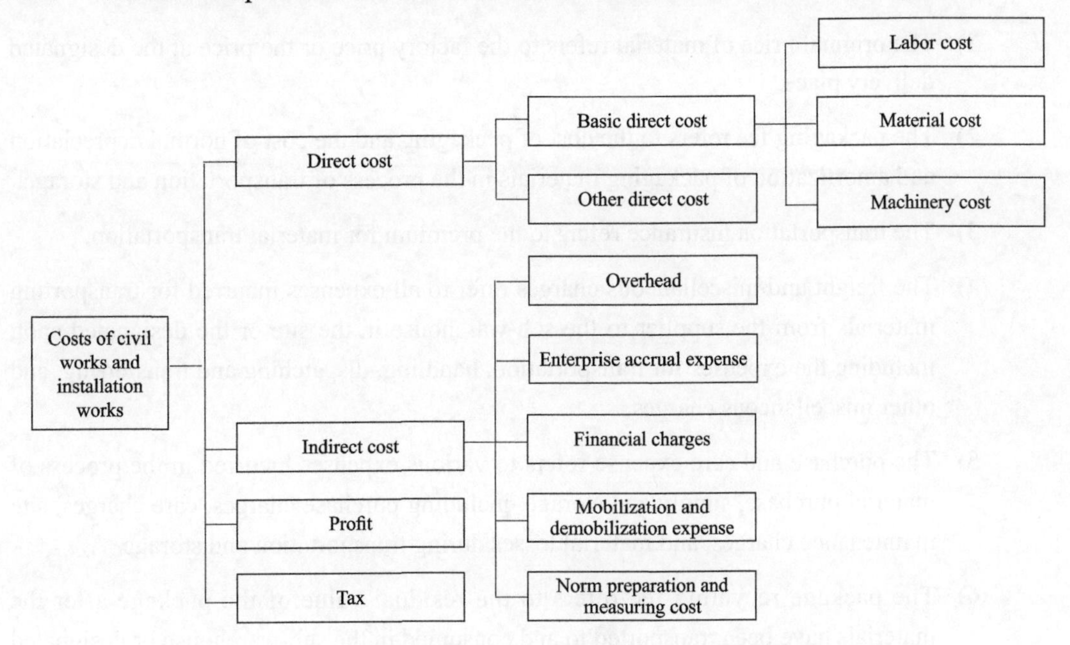

Figure 3 Cost composition of civil works and installation works

5.3.2 The costs of civil works and installation works comprise direct cost, indirect cost, profit and tax.

5.3.3 The direct cost refers to the cost directly consumed by live labor and materialized labor for civil works and installation, including basic direct cost and other direct cost.

5.3.3.1 The basic direct cost refers to the expenses in the construction process to constitute the project entity under normal construction conditions, including labor cost, material cost, and machinery cost.

a) The labor cost refers to all costs paid by the contractor to the workers engaged in civil works and installation, including basic wages, supplementary wages and social security contributions.

1) The basic wages consist of skill-based wages and post-based wages. The skill-based wages are determined according to post requirements and actual labor skills and performance. The post-based wages are determined by function and responsibility, skills, work intensity and work conditions required by a post.

2) The supplementary wages refer to the wage paid to workers in forms other than basic wages, including construction subsidies, pay for non-working days, etc. The pay for non-working days includes the wages during training, job transfer, home leave, vacation, suspension due to unfavorable climate conditions, lactation leave, sick leave for less than six months, maternity leave, wedding leave and funeral leave.

3) The social security contributions refer to the basic endowment insurance, unemployment insurance, medical insurance, maternity insurance, work-related injury insurance and housing accumulation fund that are paid to workers according to relevant national regulations and standards.

b) The material cost refers to the cost of raw materials and installation materials and amortization of reusable materials consumed in civil works and installation, including original price of material, packaging fee, transportation insurance and freight and miscellaneous charges, purchase and care expense, package recycling fee. VAT not included.

1) The original price of material refers to the factory price or the price at the designated delivery place.

2) The packaging fee refers to the cost of packaging and the cost of normal depreciation and amortization of packaging materials in the process of transportation and storage.

3) The transportation insurance refers to the premium for material transportation.

4) The freight and miscellaneous charges refer to all expenses incurred for transporting materials from the supplier to the sub-warehouse in the site or the designated spot, including the expenses for transportation, handling, dispatching and transferring, and other miscellaneous charges.

5) The purchase and care expense refers to various expenses incurred in the process of material purchase, supply and storage, including purchase charges, care charges, site maintenance charges, and material losses during transportation and storage.

6) The package recycling fee refers to the residual value of the package after the materials have been transported to and consumed in the sub-warehouse or designated spot on the site.

c) The machinery cost refers to the machinery depreciation cost, equipment repair cost, assembling /disassembling cost, operator expense, power fuel expense, insurance premium, vehicle and vessel usage tax, and machinery annual inspection fees in civil works and installation.

1) The depreciation cost refers to the amount charged off to expense of construction machinery price within the specified service period.

2) The equipment repair cost refers to the expenses of equipment and parts replacement, tools and attachment maintenance, lubricant and cleaning materials for routine maintenance, and machinery storage for the normal operation of construction machinery.

3) The assembling/disassembling cost refers to the expenses of assembling and disassembling of machinery for transportation into and out of site, testing, and the

use of auxiliary facilities.

4) The operator expense refers to the wage paid to the machinery operators.

5) The power fuel expense refers to the cost of water, electricity and fuel needed for the normal operation of construction machinery.

6) The insurance, vehicle and vessel usage tax and machinery annual inspection fee refer to the taxes and fees incurred during the use of construction machinery.

5.3.3.2 Other direct cost refers to the expenses for the completion of incorporeal projects incurred before and during construction, consisting of extra cost for construction in winter and rainy season, extra cost for construction at night, extra cost for construction in special areas, cost of construction tools, cost of temporary facilities, and others.

a) The extra cost for construction in winter and rainy season refers to the extra expenses required for continuous construction during winter and rainy season within reasonable time limit, including the expenses for heating and curing, rain proof, moisture proof, anti-slip, frost protection, and snow removal, and the expenses for extra process and lower efficiency due to the above measures.

b) The extra cost for construction at night refers to the lighting facility amortization and electricity charge incurred during construction at night.

c) The extra cost for construction in special areas refers to the expenses incurred for construction in special areas, such as extremely cold and hot areas.

d) The cost of construction tools refers to the amortization and maintenance cost of production tools and inspection and test appliances not included in the fixed assets but are indispensable for the construction and production.

e) The cost of temporary facilities refers to the expense for building, maintaining and dismantling temporary buildings, structures and other temporary facilities for normal production and living needs on the site.

f) The others refer to the expenses not mentioned above, including charges for project site resurvey (construction survey control network charges), handover, checking and testing, construction drainage, construction communication, clearing of as-built project area, and maintenance before handing over the project (including maintenance and adjustment of installed equipment). The charges for checking and testing refer to the expenses incurred in general identification and inspection of building materials, components and installed equipment, including the cost of materials and chemical supplies for testing in self-established laboratories, the cost of technological innovation, and the cost of research and test, but excluding the testing fees for novel structures or materials, or the expenses for testing materials with factory certificates, destructive tests on components, or other special tests required by the project owner.

5.3.4 The indirect cost refers to overhead, enterprise accrual expense, financial charges, mobilization and demobilization expense, and norm preparation and measuring cost, which is indispensable for the project but not directly accountable to specific products for the civil works and installation.

5.3.4.1 Overhead refers to the expense incurred in construction and management organized by the contractor, consisting of the salary and social security contributions of the managerial staff, office expense, travel expense, utilization cost of fixed assets, utilization cost of tools and equipment, insurance, taxes and additional education tax, costs of technology transfer, technology development, business entertainment, bidding, advertisement, notarization, litigation, legal counsel, audit and consulting, as well as the expenses for design of the auxiliary works, engineering drawings and data, and project design, etc. All these costs should be paid by the contractor.

5.3.4.2 The enterprise accrual expense refers to the expense paid by the contractor according to the national regulations, consisting of the staff welfare, labor protection expense, labor union expenditure, staff training expense, accident insurance to workers engaged in dangerous operation.

5.3.4.3 The financial charges refer to the cost incurred by the contractor for financing, consisting of the interest expense, net exchange loss, foreign exchange transfer fees, financial institution handling fees, guarantee fees, and other expenses during financing.

5.3.4.4 The mobilization and demobilization expense refers to the expense paid by the contractor for sending staff and construction machinery (excluding Class A machinery in the norm of machine-shift cost) into or out of the construction site.

5.3.4.5 The norm preparation and measuring cost refers to the expense paid by the contractor for providing basic data for measuring, making (revising) the standard for enterprise norm and setting the standard for sector norm.

5.3.5 The profit refers to the one that shall be counted in the costs of civil works and installation works according to the market situation of a wind farm project.

5.3.6 The tax refers to VAT that shall be counted in the costs of civil works and installation works according to national tax laws and regulations.

5.4 Other cost items

5.4.1 Composition of other cost items is shown in Figure 4.

5.4.2 The land use cost refers to the related expenses for obtaining necessary construction sites in accordance with relevant national and local laws and regulations, consisting of land requisition fee, temporary land requisition fee, compensation for attachments on the land, and site clearing expense.

5.4.3 The preliminary work cost refers to the expenses incurred in carrying out various works before the completion of pre-feasibility study report review (or before the preparation for the construction of wind power projects), including project owner management cost, expenses for setting up anemometer mast, purchasing wind measurement instruments, and measuring wind in the early stage, expenses for investigation, research and test for project planning, pre-feasibility study and the preparation of the above-mentioned design documents.

5.4.4 The project construction management cost refers to the various management expenses incurred in the process of project preparation, construction, commissioning, acceptance and handover, including the construction management cost, construction supervision cost, consulting service fee, techno-economic review fee, project quality inspection and testing fee, norm preparation and management cost, project acceptance cost, and insurance.

5.4.4.1 Construction management cost refers to the management expenses needed by the legal person of a project from the preparation to the acceptance in order to ensure normal construction. It includes cost of management equipment and appliances, personnel routine outlay, and other management expenses.

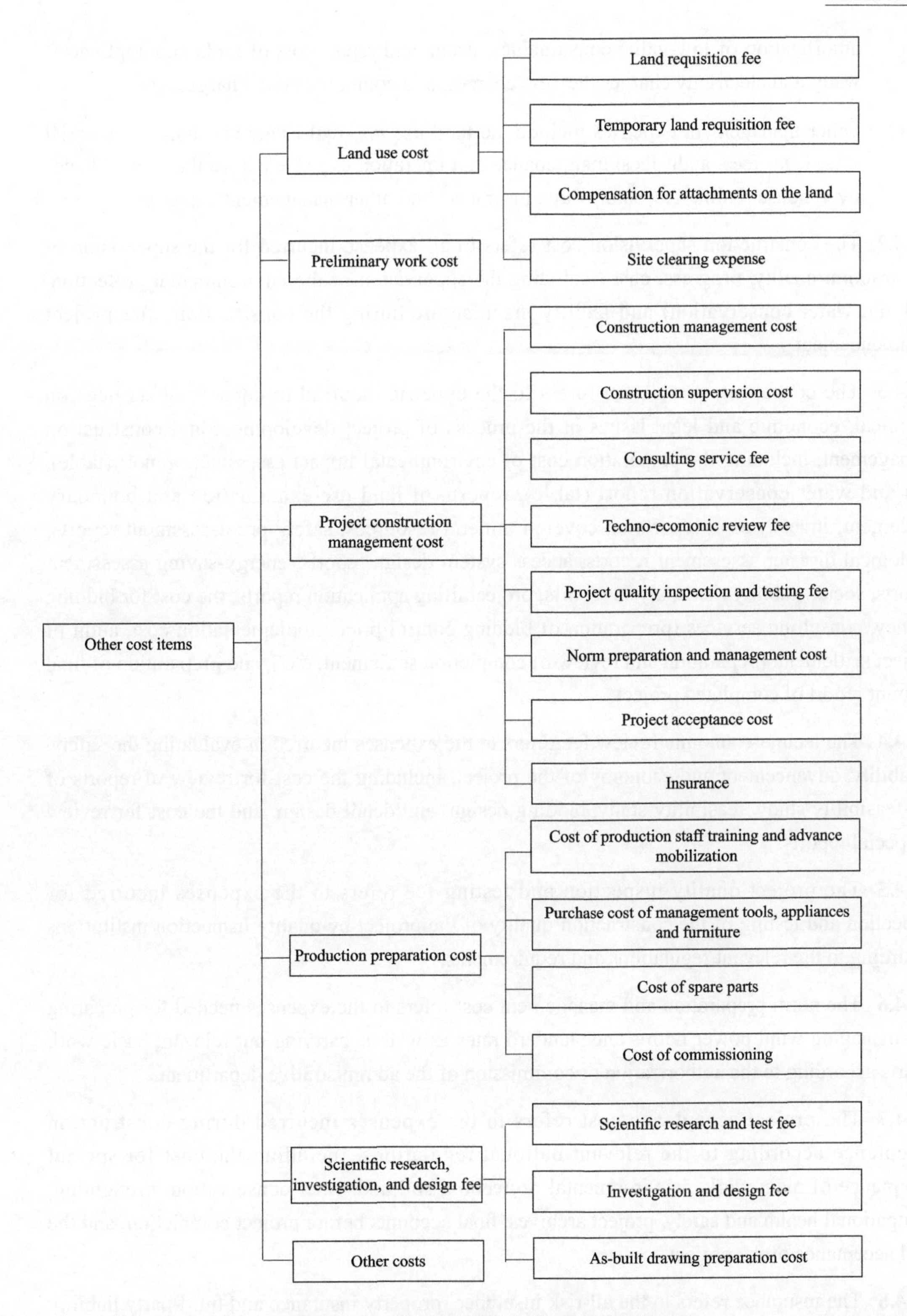

Figure 4 Composition of other cost items

a) The cost of management equipment and appliances refers to the expenses for purchasing transportation equipment, testing equipment, office and living appliances, and other equipment needed for construction management.

b) The personnel routine outlay includes personnel basic wages and supplementary wages, social security contributions, welfare, labor protection expense, personnel training expense, labor union expenditure, accident insurance to workers engaged in dangerous operation, office allowance, travel expense, conference and reception expense, expense for technical and reference books, expense for purchasing sporadic fixed assets, amortization of low-value consumables, usage and repair cost of tools and appliances, water and electricity charges, heating charges, and communication charges, etc.

c) Other management expenses include the land use tax, real estate tax, contract notarial fees, legal fees, audit fees, maintenance and operation expense before the project hand-over, house renting fee, stamp duty, insurance, and other management expenses.

5.4.4.2 The construction supervision cost refers to all expense incurred for the supervision of construction quality, progress, cost (including the supervision of the environmental protection, soil and water conservation) and facility manufacture during the construction after project commencement.

5.4.4.3 The consulting service fee refers to the expense incurred in consulting services on technical, economic and legal issues in the process of project development and construction management, including the preparation cost of environmental impact assessment report (table), soil and water conservation report (table), reports of land pre-examination and boundary settlement, investigation reports on covered mineral resources, safety pre-assessment reports, geological disaster assessment reports, access system design reports, energy-saving assessment reports, social stability risk analysis reports, project filing application reports, the cost for bidding agency, consulting services (preparation of bidding control price, implementation cost, audit of project settlement, preparation and review of completion settlement, etc.), and preparation of final account report of completed project.

5.4.4.4 The techno-economic review fee refers to the expenses incurred in evaluating the safety, reliability, advancement and economy of the project, including the cost for review of reports of pre-feasibility study, feasibility study, bidding design, and detail design, and the cost for review of special reports.

5.4.4.5 The project quality inspection and testing fee refers to the expenses incurred for inspection and testing of the construction quality of the project by quality inspection institutions according to the relevant regulations and requirements.

5.4.4.6 The norm preparation and management cost refers to the expenses needed for preparing and managing wind power norms and standard rates as well as carrying out relevant basic work efforts according to the authorization or commission of the administrative departments.

5.4.4.7 The project acceptance cost refers to the expenses incurred during construction acceptance according to the relevant national regulations, including the cost for special acceptance of main works, environmental protection, soil and water conservation, firefighting, occupational health and safety, project archives, final accounts before project completion, and the final acceptance of the project.

5.4.4.8 The insurance refers to the all-risk insurance, property insurance and third-party liability insurance for civil works, installation, and permanent equipment during project construction, which are to transfer or mitigate risks of losses due to natural disasters and unforeseen accidents.

5.4.5 The production preparation cost of a project refers to the expenses incurred due to necessary preparation by the project legal person for operation, including the cost of production staff training and advance mobilization, the purchase cost of management tools, appliances and furniture, cost of spare parts, and the cost of commissioning.

5.4.5.1 The cost of production staff training and advance mobilization into the plant consists of the expense for production staff training and advance mobilization into the plant.

a) The expense for production staff training refers to the expense incurred for training the operation, maintenance and management staff by production departments before project acceptance and commissioning, in order to ensure normal operation of the project.

b) The expense for advance mobilization of the working staff into the plant includes the staff's basic wages and supplementary wages, social security contributions, welfare, labor protection expense, staff training expense, labor union expenditure, accident insurance to workers engaged in dangerous operation, office allowance, travel expense, conference and reception expenses, expense for technical and reference books, expense for purchasing sporadic fixed assets, amortization of low-value consumables, usage and repair cost of tools and appliances, water and electricity charges, heating charges, communication charges, and other costs needed during production preparation.

5.4.5.2 The purchase cost of management tools, appliances and furniture refers to the expense for purchasing necessary office and life appliances and furniture to ensure production operation and management, excluding special tools for the equipment.

5.4.5.3 The purchase cost of spare parts refers to the expense for purchasing consumables or vulnerable spare parts and special materials needed during the installation and commissioning periods to ensure the normal production and operation of the project, excluding special tools for the equipment.

5.4.5.4 The cost of commissioning refers to the net expenses excluding the revenue of trial operation incurred during on-load commissioning of the complete set of equipment.

5.4.6 The scientific research, investigation, and design fee refers to the expense incurred in scientific research, test, investigation, and design for project construction, including scientific research and test fee, investigation and design fee, and as-built drawing preparation cost. The scientific research and test fee refers to the expense incurred in carrying out necessary scientific tests to solve technical problems in the course of project construction. The investigation and design fee refers to the expenses for investigation and design incurred in the stage of feasibility study, bidding design and detail design. The as-built drawing preparation cost refers to the cost required for the summary and compilation of drawings that can fully and truly reflect the results of the construction.

5.4.7 Others refer to the taxes and fees paid in accordance with relevant national regulations, like compensation for soil and water conservation, etc.

5.5 Contingency

5.5.1 Contingency consists of basic contingency and contingency for price variation.

5.5.2 The basic contingency refers to the expense reserved for design variations in the feasibility study design scope, measures against natural disasters, and losses caused by common natural disasters but not be compensated by insurance.

5.5.3 The contingency for price variation refers to the expense reserved for price increase due to changes of national policies and regulations, price escalation of materials and equipment, and adjustment of labor cost and other standard rates and exchange rate changes during construction.

5.6 Interest during construction

The interest during construction refers to the interest of loan fund during project construction, which is counted in the original value of fixed assets after project operation according to relevant national regulations, including bank loan interest, other loan interest and financing charges. Other financial charges refer to poundage, commitment fees, management cost and credit insurance in some debt financing.

6 Preparation of cost estimation

6.1 General requirements

6.1.1 The composition of cost estimate is shown in Figure 5.

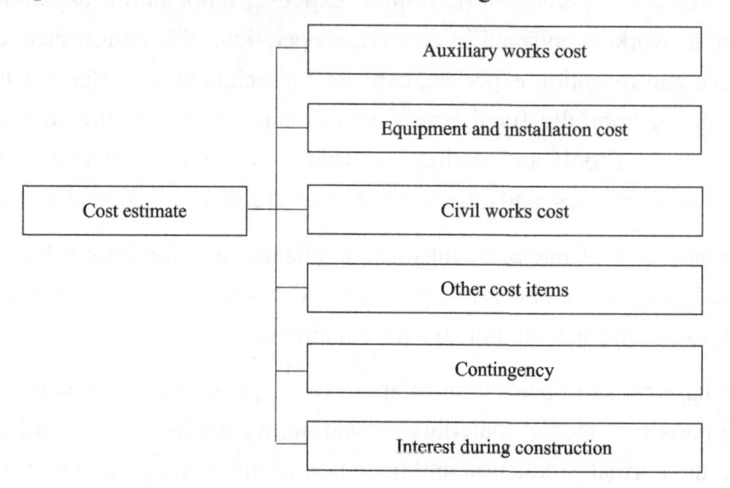

Figure 5 Composition of cost estimate

6.1.2 In addition to this standard, the preparation of cost estimate shall meet the following:

a) Relevant laws, regulations, rules and administrative normative documents issued by the state, province, autonomous region and municipalities.

b) Sector specifications and codes.

c) Norms and standard rates for cost estimate issued by energy sector norms and cost management authorities.

d) Rules for calculation of design work quantity.

e) Design documents and drawings in feasibility study stage.

6.2 Preparation of basic prices

6.2.1 Basic prices shall be prepared in accordance with relevant policies, stipulations and market price levels during cost estimation, including unit price of labor, price of material, prices of electricity and water for construction, unit price of concrete materials, and machine-shift cost.

6.2.2 The unit price of labor shall be determined according to this standard and the document issued by the energy sector and cost management authorities.

6.2.3 The price of material (excluding VAT) shall be determined by analyzing the local market price according to sources of materials and means of transportation. The calculation base of purchase and care expense is the sum of the original price, transportation insurance and freight and miscellaneous charges.

6.2.4 The price of electricity for construction is composed of basic price, amortization of electric energy loss and maintenance charge of power supply facilities:

a) There are two forms of basic price: the basic price of purchased electricity and the basic price of self-generated electricity. The basic price of purchased electricity refers to sum of the electricity price and surcharge specified by the state, province, autonomous region or municipality where the project is located, which shall be paid to the power supplier.

The basic price of self-generated electricity refers to the unit generating cost of the equipment.

b) The amortization of electricity energy loss refers to the amortization of energy losses incurred on all transformers, distribution equipment and transmission lines, from the connecting points for purchased electricity (or outgoing line side of generating equipment for self-generated electricity) to the low voltage (LV) side of the terminal step-down transformer at each construction site.

c) The amortization of maintenance cost of power supply facilities refers to the cost included in the electricity price, including basic depreciation cost, repair cost and assembling/disassembling cost of power transformation and distribution equipment, as well as the operation and maintenance cost for equipment, transmission and distribution lines.

d) The price of electricity from power grid shall be calculated by the following formula:

$$E_1 = \frac{B}{(1-L_1)(1-L_2)} + M \quad (1)$$

where

E_1 is the price of electricity from power gird [CNY/(kW · h)];

B is the basic price (excluding VAT) [CNY/(kW · h)];

L_1 is the power loss rate of HV transmission line, taken as 2 % to 5 %;

L_2 is the power loss rate of transformer and distribution equipment and distribution lines, taken as 4 % to 6 %;

M is the amortization of maintenance cost for power supply facilities (excluding VAT), taken as 0.03 CNY/(kW · h).

e) The price of electricity from diesel generator shall be calculated by the following formula:

$$E_2 = \frac{C}{(P \times H \times K_1 \times K_2)(1-L_3)} + M \quad (2)$$

where

E_2 is the price of electricity from diesel generator [CNY/(kW · h)];

C is the total machine-shift cost of diesel generator [CNY/(kW · h)];

P is the sum of rated capacity of diesel generator [CNY/(kW · h)];

H is the machine-shift hourage, taken as 8;

K_1 is the time utilization coefficient, taken as 0.7 to 0.8;

K_2 is the output coefficient of diesel generator, taken as 0.8;

L_3 is the power loss rate of transformer and distribution equipment and distribution lines, taken as 2 % to 3 %;

M is amortization of maintenance cost for power supply facilities (excluding VAT), taken as 0.03 CNY/(kW · h).

According to the construction planning of the wind farm, the electricity price shall be calculated by the weighted average method considering the electric quantity and electricity prices from different means of power supply.

6.2.5 The price of water for construction shall be calculated according to the water supply mode determined by the construction planning.

6.2.6 Unit price of raw materials of concrete:

a) The unit price of self-provided raw materials of concrete, according to the basis of strength, gradations, ages and anti-freezing indexes, shall be the unit prices of the cement, mineral admixtures, aggregates, chemical admixtures, and water for per cubic meter of concrete, which shall be included in the unit price of concrete.

b) The unit price of commercial concrete (excluding VAT) shall be determined by both the ex-factory price and the freight charge.

6.2.7 The calculation of machine-shift cost shall comply with NB/T 31010-2019.

6.3 Preparation of unit price for civil works and installation works

6.3.1 The calculation of unit price of civil works shall be as specified in Table 1.

Table 1 Calculation of unit price of civil works

No.	Item	Calculation
1	Direct cost	Basic direct cost + Other direct cost
1.1	Basic direct cost	Labor cost + Material cost + Machinery cost
1.1.1	Labor cost	∑ (Labor quantity as per norm × Unit price of labor)
1.1.2	Material cost	∑ (Material quantity as per norm × Unit price of material)
1.1.3	Machinery cost	∑ (Machine quantity as per norm × Machine-shift cost)
1.2	Other direct cost	(Labor cost + Machinery cost) × Other direct cost rate
2	Indirect cost	(Labor cost + Machinery cost) × Indirect cost rate
3	Profit	(Labor cost + Machinery cost + Other direct cost + Indirect cost) × Profit rate
4	Tax	(Direct cost + Indirect cost + Profit) × VAT rate
5	Unit price for civil works	Direct cost + Indirect cost + Profit + Tax

6.3.2 The calculation of unit price of installation based on consumption shall be as specified in Table 2.

Table 2 Calculation of unit price of installation based on consumption

No.	Item	Calculation
1	Direct cost	Basic direct cost + Other direct cost
1.1	Basic direct cost	Labor cost + Material cost + Machinery cost + Installation material cost
1.1.1	Labor cost	∑(Labor quantity as per norm × Unit price of labor)
1.1.2	Material cost	∑ (Material quantity as per norm × Unit price of material)
1.1.3	Machinery cost	∑ (Material quantity as per norm × Unit price of material)

Table 2 *(continued)*

No.	Item	Calculation
1.1.4	Installation material cost	Installation material quantity not listed in norms × Unit price of material
1.2	Other direct cost	(Labor cost + Machinery cost) × Other direct cost rate
2	Indirect cost	Labor cost × Indirect cost rate
3	Profit	(Labor cost + Machinery cost + Other direct cost + Indirect cost) × Profit rate
4	Tax	(Direct cost + Indirect cost + Profit) × VAT rate
5	Unit price of installation	Direct cost + Indirect cost + Profit + Tax
NOTE	Installation materials do not include those provided by the owner.	

6.3.3 The calculation of unit price of installation based on rate or price list shall be as specified in Table 3.

Table 3 Calculation of unit price of installation based on rate or price list

No.	Item	Calculation
1	Direct cost	Basic direct cost + Other direct cost
1.1	Basic direct cost	Labor cost + Material cost + Machinery cost + Installation material cost
1.1.1	Labor cost	Labor cost as per norm
1.1.2	Material cost	Material cost as per norm
1.1.3	Machinery cost	Machinery expense as per norm
1.1.4	Installation material cost	Installation material cost as per norm
1.2	Other direct cost	(Labor cost + Machinery cost) × Other direct cost rate
2	Indirect cost	Labor cost ×Indirect cost rate
3	Profit	(Labor cost + Machinery cost + Other direct cost + Indirect cost) × Profit rate
4	Tax	(Direct cost + Indirect cost +Profit) × VAT rate
5	Unit price of installation	Direct cost + Indirect cost + Profit + Tax

6.4 Preparation of equipment cost

6.4.1 Equipment cost includes the original price, freight and miscellaneous charges, transportation insurance, purchase and care expense.

6.4.2 For domestic equipment, the factory price is taken as the original price, which may be determined based on the price offered by the manufacturer and market price level. For imported equipment, the original price is the sum of the CIF, tariff, VAT, banking charge, inspection fee, and port charge.

6.4.3 Freight and miscellaneous charges shall be calculated by multiplying the original price of the equipment by the freight and miscellaneous charge rate.

6.4.4 The transportation insurance shall be calculated by multiplying the original equipment price by the transportation insurance rate.

6.4.5 Purchase and care expense shall be calculated as a percentage of the sum of the original equipment price, freight and miscellaneous charges, and transportation insurance.

6.5 Preparation of cost estimate

6.5.1 Auxiliary works

6.5.1.1 The cost for temporary road works shall be calculated by multiplying the design work quantities by the unit price, or by the local unit cost index.

6.5.1.2 The cost of power supply for construction shall be calculated by multiplying the design work quantity by the unit price, or by the local unit cost index. The cost includes the power supply lines of 10 kV and above connected from the power grid to the site and the power supply facilities of 35 kV and above. However, it does not include the maintenance cost of power supply lines, power transformers or distribution facilities, which shall be included in the cost of power supply for construction as amortization.

6.5.1.3 The cost of the crane hardstand for wind turbine installation shall be calculated by multiplying the design work quantity by the unit price.

6.5.1.4 The cost of other auxiliary works shall be calculated according to the construction planning and the project-specific conditions.

6.5.1.5 The cost of safe and civilized construction measures shall be calculated as a percentage of civil works cost and installation cost (excluding the cost calculated by the unit cost index or the cost of safe and civilized construction measures).

6.5.2 Equipment and installation works

6.5.2.1 The cost estimate of equipment and installation works shall be prepared respectively for equipment and installation.

6.5.2.2 The equipment cost shall be calculated by multiplying the equipment quantity by the unit price .

6.5.2.3 The cost of vehicles for production and maintenance shall be calculated by multiplying the number of vehicles by the unit price.

6.5.2.4 The amortization of equipment in control center shall be determined by the project owner's planning scheme.

6.5.2.5 The installation cost can be calculated in the following two ways:

a) The installation cost is calculated by multiplying the equipment quantity by the unit price of installation when the unit price is in the form of consumption or price list. The installation materials provided by the project owner are listed separately in the cost estimate of equipment and installation at the price including tax.

b) The installation cost is calculated by multiplying the original price of the equipment by installation rate when in the form of rate.

6.5.3 Civil works

6.5.3.1 The cost of wind farm works and collection line works are calculated by multiplying the work quantity by the unit price.

6.5.3.2 The cost of step-up substation works shall be calculated by multiplying the work quantity by the unit price or by the unit cost indexes.

a) The costs of site grading, foundation works of main transformer, foundation works of reactive power compensators, foundation works of power distribution equipment, and of structures of power distribution equipment shall be calculated by multiplying the work quantity by the unit price.

b) The costs of production buildings, auxiliary production buildings, site offices and living quarters shall be calculated by multiplying the floor area by the unit cost index. The floor area of site buildings shall be determined by design, and the unit cost index shall be calculated according to the local unit cost indexes of the building works and related data. Work breakdown may be adjusted according to actual design scheme.

c) The cost of outdoor works includes wall, gate, plantation, roads, ground hardening, water supply pipe, drainage pipe, inspection well, rainwater well, sewage well, manhole cover, valves, septic tank, drainage ditch, etc. Among them, the wall, gate, plantation, roads and ground hardening shall be calculated separately, and the costs of other outdoor works shall be calculated as a percentage of the sum of production buildings, auxiliary production buildings, offices and living quarters, and outdoor works (excluding other outdoor works).

6.5.3.3 The cost of roads shall be calculated by multiplying the work quantity by the unit price , or by the local unit cost index.

6.5.3.4 Costs of other works:

a) The costs of environmental protection works, soil and water conservation works, and occupational health and safety works shall be calculated according to specialized design results.

b) The costs of safety monitoring works, water supply for firefighting, production and living, and flood (tide) control works shall be calculated by multiplying the design work quantity by the unit price.

c) The amortization of management facilities for centralized production and operation shall be determined by the project owner's planning scheme.

6.5.4 Other cost items

6.5.4.1 The land use cost shall be calculated according to the land area determined by design and the standards issued by province-level government.

6.5.4.2 Project owner management cost, expenses for setting anemometer tower, purchasing wind measuring equipment, and measuring wind in preliminary work cost may be listed and calculated according to the actual situation. The planning cost may be calculated according to the actual cost and the total installed capacity of the planned wind farm. The pre-feasibility study cost may be calculated according to the calculation standard of the investigation and design cost.

6.5.4.3 Project construction management costs are calculated as follows:

a) The construction management cost is calculated as a percentage of costs of civil works and installation works.

b) The construction supervision cost is calculated as a percentage of costs of civil works and installation works.

c) The consulting service fee is calculated as a percentage of costs of civil works and installation works.

d) The techno-economic review cost is calculated as a percentage of costs of civil works and installation works.

e) The project quality inspection and testing fee is calculated as a percentage of civil works and installation cost.

f) The norm preparation and management cost is calculated as a percentage of costs of civil works and installation works.

g) The project acceptance cost is calculated as a percentage of costs of civil works and installation works.

h) The construction insurance is calculated as a percentage of the sum of costs of civil works and installation works and equipment cost.

6.5.4.4 Production preparation costs are calculated as follows:

a) The costs of production staff training and advance mobilization are calculated as a percentage of costs of civil works and installation works.

b) The purchase cost of management tools, appliances and furniture is calculated as a percentage of costs of civil works and installation works.

c) The purchase cost of spare parts is calculated as a percentage of costs of civil works and installation works. When spare parts are included in the expense of wind turbines, the corresponding equipment cost shall be deducted from the calculation base.

d) The cost of commissioning is calculated as a percentage of costs of civil works and installation works, but the revenue from power generation during commissioning period shall be deducted.

6.5.4.5 The scientific research, investigation and design fee is calculated as follows:

a) The costs of scientific research and test are calculated as a percentage of costs of civil works and installation works.

b) The investigation and design of a wind farm is divided into five stages: planning stage, pre-feasibility study stage, feasibility study stage, bidding design stage and detail design stage. The cost for planning and pre-feasibility study is listed in the preliminary work cost. The pre-feasibility study cost is calculated as a percentage of the sum of the investigation and design cost incurred in stages of feasibility study, bidding design, and detail design. The investigation and design cost refers to the investigation expense and design expense incurred in stages of feasibility study, bidding design, detail design, which is calculated according to investigation cost rates and design cost rates respectively on civil works and installation calculation base.

c) The as-built drawing preparation cost is calculated as a percentage of the sum of costs incurred in feasibility study, bidding design and detail design.

6.5.4.6 Other taxes and fees are calculated according to relevant documents issued by the state, province, autonomous region, and/or municipality.

6.5.5 Contingency

6.5.5.1 The basic contingency is calculated as a percentage of the sum of auxiliary works cost, equipment and installation cost, civil works cost, and other cost items.

6.5.5.2 The contingency for price variation shall be calculated on the basis of annual cost (including basic contingency) according to the construction period. It shall be calculated from the next year after the price level year in the cost estimation. The annual price index of wind power project is determined according to energy sector norms and price index issued by cost management authorities.

6.5.6 Interest during construction

6.5.6.1 The interest during construction shall be calculated based on annual cost (deducting capital) starting from the start-up period, according to cost amount, fund sources and arrangement. The benchmark loan interest rates of five-year period or above issued by the People's Bank of China during the cost estimate preparation period shall be applied for bank loan.

6.5.6.2 The loan interest incurred before the first group (batch) of wind turbine generators putting into operation is included in the project construction cost. The loan interest incurred afterwards shall be divided according to the production capacity and included in capital construction cost and operation production cost respectively.

6.5.7 Preparation of total investment

6.5.7.1 The static investment of a project is the sum of auxiliary works cost, equipment and installation cost, civil works cost, other cost items, and basic contingency.

6.5.7.2 The total investment of a project is the sum of static investment, contingency for price variation, and interest during construction.

6.5.7.3 If a project includes power transmission works, the cost of power transmission is not included in the total investment of the wind power project and is listed separately.

7 Standard rates

7.1 Standard rates for unit price of civil works and installation works

7.1.1 The unit price of labor is calculated according to Table 4.

Table 4 Unit price of labor

No.	Skill level of labor	Wage standard (CNY/man-day)
1	Highly skilled	249
2	Skilled	173
3	Unskilled	120

7.1.2 The rate of purchase and care expense is 2.8 %, and the calculation base is the sum of the original costs of material, transportation insurance, and freight and miscellaneous charges.

7.1.3 Other direct costs are calculated according to the following provisions respectively:

a) The rate of extra cost for construction in winter and rainy season is given in Table 5, and the regional classification is shown in Table 6.

b) The rate of extra cost for construction at night is given in Table 7.

c) The rate of extra cost for construction in special areas is given in Table 8.

d) The rate of usage fee for construction tools and appliances is given in Table 9.

Table 5 Rate of extra cost for construction in winter and rainy season

Category of works		Calculation base	Regional classification				
			I	II	III	IV	V
			Rate (%)				
Civil works		Labor cost and machinery cost	0.45	0.64	0.95	1.37	1.71
Installation works	Wind turbines and tower		0.35	0.49	0.75	1.07	1.33
	Collection lines		1.11	1.56	2.36	3.11	3.47
	Other equipment		1.27	1.80	2.73	3.61	4.02

Table 6 Regional classification

Regional classification	Province, autonomous region and municipality
I	Shanghai, Jiangsu, Anhui, Fujian, Hubei, Hunan, Guangdong, Guangxi, Hainan, Jiangxi, Zhejiang
II	Beijing, Tianjin, Shandong, Henan, Hebei (south of Chengde in Zhangjiakou), Chongqing, Sichuan (except Garze and Aba), Yunnan (except Diqing), Guizhou
III	Liaoning (Gaizhou and south of Gaizhou), Shaanxi (south of Yulin), Shanxi, Hebei (Zhangjiakou, Chengde and their north areas)
IV	Liaoning (north of Gaizhou), Shaanxi (Yulin and north of Yulin), Inner Mongolia (Xilin Gol League; leagues, cities and counties south of Xilinhot city, (excluding Alxa League), Jilin, Gansu, Ningxia, Sichuan (Garze and Aba), Yunnan (Diqing), Xinjiang (south of Yili and Hami)
V	Heilongjiang, Qinghai, Xinjiang (Yili, Hami and their north areas), other areas in Inner Mongolia than the areas mentioned in IV

Table 7 Rate of extra cost for construction at night

Category of works		Calculation base	Rate (%)
Civil works		Labor cost and machinery cost	0.11
Installation works	Wind turbines and tower		0.06
	Collection lines		0.08
	Other equipment		0.17

Table 8 Rate of extra cost for construction in special areas

Category of works		Calculation base	High-latitude cold regions (%)	Severe hot regions (%)
Civil works		Labor cost and machinery cost	1.96	1.72
Installation works	Wind turbines and tower		1.96	1.72
	Collection lines		1.96	1.72
	Other projects		2.61	2.29
NOTE 1 High-latitude cold regions refer to the regions at 45 ° N and above. NOTE 2 Severe hot regions refer to Turpan Prefecture of Xinjiang.				

Table 9 Rate of cost for construction tools and appliances

Category of works		Calculation base	Rate (%)
Civil works		Labor cost and machinery cost	1.34
Installation works	Wind turbines and tower		0.67
	Collection lines		2.63
	Other equipment		0.74

e) The rate of temporary facility fee is given in Table 10, and regional classification is shown in Table 6.

Table 10 Rate of cost for temporary facility

Category of works		Calculation base	Regional classification				
			Ⅰ	Ⅱ	Ⅲ	Ⅳ	Ⅴ
			Rate (%)				
Civil works		Labor cost and machinery cost	4.28	5.18	5.92	6.28	6.68
Installation works	Wind turbines and tower		1.73	1.92	2.00	2.08	2.20
	Collection lines		2.15	2.39	2.49	2.59	2.73
	Other equipment		0.63	0.70	0.73	0.76	0.80

f) The rate of other costs is given in Table 11.

Table 11 Rate of other costs

Category of works		Calculation base	Rate (%)
Civil works		Labor cost and machinery cost	1.86
Installation works	Wind turbines and tower		1.73
	Collection lines		2.40
	Other equipment		2.30

7.1.4 The rate of indirect cost is given in Table 12.

Table 12 Rate of indirect cost

Category of works	Calculation base	Rate of indirect cost (%)					
		Total	Overhead	Enterprise accrual expense	Financial charge	Mobilization and demobilization expenses	Norm determination and preparation costs
Civil works	Labor cost and machinery cost	27.66	17.48	5.73	3.65	0.60	0.20
Installation works	Labor cost	74.00	45.33	10.87	10.26	5.66	1.88

7.1.5 The profit rate is 10 %, and the calculation base is the sum of labor cost, machinery cost, and other direct and indirect costs.

7.1.6 The tax rate is 9 %, and the calculation base is the sum of direct cost, indirect cost and profit.

7.2 Standard rates of equipment costs

7.2.1 The rate of freight and miscellaneous charges is taken as 2 % to 3 % for major equipment (wind turbines, towers and main transformers) and 3 % to 5 % for other equipment. If the main equipment is directly delivered to the site, only unloading fee is charged, which is calculated as 0.1 % of the transportation expense. For possible transfer of wind turbines and towers, the expenses may be estimated according to the design scheme.

7.2.2 The rate of equipment transportation insurance is 0.4 % of the original price of the equipment.

7.2.3 The rate of equipment purchase and care expense is 0.5 % of the sum of the equipment original price, freight and miscellaneous charge and transportation insurance.

7.3 Standard rates of auxiliary works and civil works

7.3.1 The rate of cost for safe and civilized construction measures is 2 %, and the calculation base is the costs of civil works and installation works (excluding the project investment calculated by the unit cost index and the cost of safe and civilized construction measures).

7.3.2 The rate of cost for other outdoor works is 5 % to 10 % , and the calculation base is the sum of the costs for production buildings, auxiliary production buildings, site offices and living quarters, and outdoor works (excluding other outdoor works).

7.4 Standard rates of other cost items

7.4.1 The rate of construction management cost is given in Table 13.

Table 13 Rate of construction management cost

No.	Amount (million CNY)	Calculation base	Rate (%)
1	≤ 50	Costs of civil works and installation works	7.98
2	100		4.71
3	200		3.11

Table 13 *(continued)*

No.	Amount (million CNY)	Calculation base	Rate (%)
4	300	Costs of civil works and installation works	2.41
5	400		1.97
6	≥ 600		1.54
NOTE For the amount between the above adjacent figures, the rate is determined by the interpolation method.			

7.4.2 The rate of construction supervision cost is given in Table 14.

Table 14 Rate of construction supervision cost

No.	Amount (million CNY)	Calculation base	Rate (%)
1	≤ 50	Costs of civil works and installation works	3.18
2	100		2.54
3	200		1.83
4	300		1.51
5	400		1.32
6	≥ 600		1.18
NOTE For the amount between the above adjacent figures, the rate is determined by the interpolation method.			

7.4.3 The rate of consulting service fee is given in Table 15.

Table 15 Rate of consulting fee

No.	Amount (million CNY)	Calculation base	Rate (%)
1	≤ 50	Costs of civil works and installation works	6.06
2	100		3.56
3	200		1.82
4	300		1.47
5	400		1.21
6	≥ 600		0.95
NOTE For the amount between the above adjacent figures, the rate is determined by the interpolation method.			

7.4.4 The rate of techno-economic review fee is given in Table 16.

7.4.5 The rate of project quality inspection and testing fee is 0.2 %, and the calculation base is the costs of civil works and installation works.

7.4.6 The rate of norm preparation and management cost is 0.15 %, and the calculation base is of the costs of civil works and installation works.

7.4.7 The rate of project acceptance cost is given in Table 17.

Table 16 Rate of techno-economic review fee

No.	Amount (million CNY)	Calculation base	Rate (%)
1	≤ 50	Costs of civil works and installation works	1.68
2	100		1.05
3	200		0.60
4	300		0.47
5	400		0.39
6	≥ 600		0.30
NOTE For the amount between the above adjacent figures, the rate is determined by the interpolation method.			

Table 17 Rate of project acceptance cost

No.	Amount (million CNY)	Calculation base	Rate (%)
1	≤ 50	Costs of civil works and installation works	2.38
2	100		1.33
3	200		0.86
4	300		0.55
5	400		0.41
6	≥ 600		0.28
NOTE For the amount between the above adjacent figures, the rate is determined by the interpolation method.			

7.4.8 The rate of construction insurance is 0.3 %, and the calculation base is the sum of costs of civil works and installation works and the purchase cost of equipment.

7.4.9 Production preparation cost is calculated according to the following items:

a) The rate of production staff training and advance mobilization cost is given in Table 18.

Table 18 Rate of costs for production staff training and advance mobilization

No.	Amount (million CNY)	Calculation base	Rate (%)
1	≤ 50	Costs of civil works and installation works	1.11
2	100		0.84
3	200		0.63
4	300		0.52
5	400		0.43
6	≥ 600		0.35
NOTE For the amount between the above adjacent figures, the rate is determined by the interpolation method.			

b) The rate of purchase cost for tools, appliances and furniture for production management is given in Table 19.

Table 19 Rate of purchase cost for tools, appliances and furniture for production management

No.	Amount (million CNY)	Calculation base	Rate (%)
1	≤ 50	Costs of civil works and installation works	2.23
2	100		1.67
3	200		1.26
4	300		1.05
5	400		0.87
6	≥ 600		0.71
NOTE For the amount between the above adjacent figures, the rate is determined by the interpolation method.			

c) The rate of purchase cost for spare parts is 0.3 %, and the calculation base is the equipment cost.

d) The rate of joint commissioning cost is 0.4 %, and the calculation base is the installation works cost. The revenue of power generation during the commissioning period is calculated as follows:

$$A = S \times T \times E_3 \times K_3 \tag{3}$$

where

A is the revenue of power generation during the commissioning period (CNY);

S is the installed capacity of the project (10^4 kW);

T is the time of commissioning (h), taken as 240;

E_3 is the desulfuration benchmark on-grid price of coal-fired power plant where the project is located [CNY/(kW · h)];

K_3 is the comprehensive coefficient of the power generation output, ranging from 0.5 to 0.8 according to the project commissioning time and annual time distribution of wind energy resources in the wind farm.

7.4.10 The rate of scientific research and test cost is 0.50 %, and the calculation base is the costs of civil works and installation works.

7.4.11 Investigation and design cost includes investigation cost and design cost, which shall be calculated according to the following formulae, respectively:

$$F_1 = D \times R_1 \tag{4}$$

$$F_2 = D \times R_2 \tag{5}$$

where

F_1 is the investigation cost (CNY);

F_2 is design cost (CNY);

D is the cost of civil works and installation works (CNY);

R_1 is the rate of investigation cost (%);

R_2 is the rate of design cost (%).

In the above formulae, the rates of investigation cost and design cost are given in Tables 20 and 21, respectively. The project ranking and the score value of investigation and design complexity are given in Tables 22 and 23.

Table 20 Rate of investigation cost

No.	Total installed capacity (MW)	Project ranking		
		Class Ⅰ	Class Ⅱ	Class Ⅲ
		Rates (%)		
1	30	1.23	1.36	1.51
2	50	0.81	0.89	0.98
3	100	0.58	0.65	0.71
4	200	0.50	0.56	0.61
5	300	0.44	0.49	0.54
6	400	0.38	0.42	0.46
7	500	0.31	0.35	0.38
NOTE When the total installed capacity is between two adjacent items, the interpolation method is used.				

Table 21 Rate of design cost

No.	Total installed capacity (MW)	Project ranking		
		Class Ⅰ	Class Ⅱ	Class Ⅲ
		Rates (%)		
1	30	4.90	5.45	5.99
2	50	3.20	3.56	3.92
3	100	2.33	2.58	2.84
4	200	2.00	2.22	2.44
5	300	1.78	1.97	2.17
6	400	1.50	1.67	1.83
7	500	1.26	1.40	1.54
NOTE When the total installed capacity is between two adjacent items, the interpolation method is used.				

Table 22 Project ranking

Ranking	Scoring
Class Ⅰ	The sum of the score value of design complexity is no more than 5
Class Ⅱ	The sum of the score value of design complexity is 5 to 10
Class Ⅲ	The sum of the score value of design complexity is no less than 10

Table 23 Score value of investigation and design complexity

Item	Project design conditions	Score value
Tower	Hub height ≤ 80 m	–2
	Hub height 80 m - 100 m	1
	Hub height > 100 m	2
Foundation treatment of wind power turbine unit	Simple: good geological conditions without foundation treatment	–2
	Medium: conventional foundation treatment	2
	Complex: complex geological conditions requiring special foundation treatment	6
Environmental protection	Simple environmental requirements	–2
	General environmental requirements	1
	Special environmental requirements	2
Collection lines	All cables buried	0
	Mainly overhead lines	3
Site roads	Simple: conventional design with small work quantity	0
	Medium: moderate design complexity and work quantity	2
	High design complexity with large work quantity	6
Step-up substation	Voltage class ≤ 110 kV	–2
	Voltage class = 220 kV	1
	Voltage class ≥ 330 kV	2
Terrain	Plains with low topographic relief or relative altitude ≤ 20 m	–2
	Hilly lands/uplands with high topographic relief or 20 m < relative altitude ≤ 80 m	2
	Mountainous regions with extremely high topographic relief or relative altitude > 80 m	6
	River benchland or tidal flats	4

7.4.12 The proportion of investigation and design fee in different stages is given in Table 24.

Table 24 Proportion of investigation and design fee in different stages

Item	Design stages		
	Feasibility study	Bidding design	Detail design
	Proportion (%)		
Investigation fee	15	65	20
Design fee	20	35	45

7.4.13 The rate of pre-feasibility study cost is 12 %, and the calculation base is the sum of investigation and design fee in feasibility study, bidding design and detail design stages.

7.4.14 The rate of preparation cost of as-built drawing is 8 %, and the calculation base is the sum of the design fees in feasibility study, bidding design and detail design stages.

7.5 Standard rates of contingency

7.5.1 The rate of basic contingency is 1 % to 3 %, and the calculation base is the sum of the auxiliary works cost, equipment and installation works cost, civil works cost, and other cost items.

7.5.2 The annual price index of contingency for price variation is provisionally 0.

8 Documentation of cost estimation

8.1 The documentation of cost estimation of a wind power project comprises the front cover, signed and stamped title page, preparation explanation, tables of cost estimate, and attached tables.

8.2 The signed and stamped title page shall be in the format shown in Figure B.1 of Annex B. The main drafting, checking and reviewing personnel shall put their signatures and the special seal of registered cost engineer on this page.

8.3 The preparation explanation shall include the project overview, preparation principles and basis, basic prices, standard rates, cost estimation of various works, other issues need to be explained, and the table of main techno-economic indicators.

8.3.1 The project overview shall summarize the project location and scale, access and transportation conditions, main work quantities, construction period, relevant natural geographical conditions, geological and geomorphological situations, fund resources and capital percentage, and present such major indicators as the total investment, static investment, costs per kW and per kW · h, etc.

8.3.2 The preparation principles and basis shall present the relevant standards and regulations, norms and standard rates, design documents, and the price level during the preparation of cost estimation.

8.3.3 The basic prices shall present the basis and results of unit price of labor, main material prices, main equipment prices and other basic prices.

8.3.4 The standard rates shall present the rates used for calculating the unit prices of equipment installation and civil works.

8.3.5 The cost estimation of various works shall present the methods for estimating the auxiliary works cost, equipment and installation cost, civil works cost, other cost items, contingency, and the interest during construction.

8.3.6 Other issues need to be explained refer to the issues that need to be explained besides the above- mentioned contents in the cost estimation.

8.3.7 The table of main techno-economic indicators shall be in the format shown in Table B.1 of Annex B.

8.4 The tables of cost estimate include summary cost estimate, cost estimate of auxiliary works, cost estimate of equipment and installation works, cost estimate of civil works, cost estimate of other cost items, and cost estimate by year. All the tables shall be in the formats shown in Table B.2 to Table B.7 of Annex B.

8.5 The attached tables include summary of unit price of installation works, summary of unit price of civil works, summary of machine-shift cost, unit price of raw materials of concrete, unit price of civil works, and unit price of installation works. The tables shall be in the formats shown in Table B.8 to Table B.13 of Annex B.

Annex A
(normative)
Work breakdown

Table A.1 to Table A.4 show the breakdowns of cost estimation for onshore wind power projects.

Table A.1 shows the breakdown of auxiliary works.

Table A.1 Breakdown of auxiliary works

No.	Level 1	Level 2	Level 3	Techno-economic indicator
I	Temporary road works			
1		Highway		
			Earth excavation	CNY/m^3
			Rock excavation	CNY/m^3
			Earth and rock backfill	CNY/m^3
			Masonry	CNY/ m^3
			Concrete	CNY/m^3
			Steel bar fabrication and installation	CNY/t
			Subgrade and cushion	CNY/m^2
			Surface	CNY/m^2
			Auxiliary facilities	CNY/m
2		Bridge (culvert)		
			Earth excavation	CNY/m^3
			Rock excavation	CNY/m^3
			Earth and rock backfill	CNY/m^3
			Masonry	CNY/m^3
			Concrete	CNY/m^3
			Steel bar fabrication and installation	CNY/t
			Steel structure fabrication and installation	CNY/t
			Pavement	CNY/m^2
			Culvert	CNY/m
			Auxiliary facilities	CNY/m
II	Power supply for construction			

Table A.1 *(continued)*

No.	Level 1	Level 2	Level 3	Techno-economic indicator
1		Power supply lines		CNY/km
2		Power supply facilities		CNY/set
Ⅲ	Crane hardstand for wind turbine installation			
			Site grading	CNY/m²
			Earth excavation	CNY/m³
			Rock excavation	CNY/m³
			Earth and rock backfill	CNY/m³
			Masonry	CNY/m³
Ⅳ	Other auxiliary works			
1		Mobilization and demobilization of large hoisting machinery		CNY/set
2		Water supply		CNY/lot
3		Cofferdam		CNY/lot
4		Site grading for temporary facilities of wind farm in mountainous areas		CNY/lot
V	Safe and civilized construction measures			CNY/lot
The temporary auxiliary works for permanent use shall be listed under the corresponding permanent works.				

Table A.2 shows the breakdown of equipment and installation.

Table A.2 Breakdown of equipment and installation

No.	Level 1	Level 2	Level 3	Techno-economic indicator
I	Power generation equipment and installation			
1		Wind turbine		
			Wind turbine generator	CNY/set
2		Tower		
			Tower	CNY/set
			Foundation ring	CNY/group
			Prestressed anchor bolt	CNY/group
3		Wind turbine outgoing lines		

Table A.2 *(continued)*

No.	Level 1	Level 2	Level 3	Techno-economic indicator
			Cable laying	CNY/km
			Cable terminal	CNY/set
4		Transformer		
			Box-type substation	CNY/set
			Other transformers	CNY/set
			Fuse	CNY/piece
			Circuit breaker	CNY/set
			Isolating switch	CNY/set
			Load switch	CNY/set
			Lightning arrester	CNY/set
5		Earthing		
			Earthing bus	CNY/t or CNY/m
			Earthing electrode	CNY/piece
			Resistance-reducing agent	CNY/t
Ⅱ	Collection line equipment and installation			
1		Power collection cable		
			Cable laying	CNY/km
			Optical cable laying	CNY/km
			Cable intermediate joint	CNY/set
			Terminal	CNY/set
			Cable tray (bracket)	CNY/t or CNY/m
			Cable distribution box	CNY/piece
2		Overhead collection lines		
			Cement pole assembling and erection	CNY/pole
			Steel pole assembling and erection	CNY/pole
			Iron tower assembling and erection	CNY/pole
			Fuse	CNY/piece
			Circuit breaker	CNY/set

Table A.2 *(continued)*

No.	Level 1	Level 2	Level 3	Techno-economic indicator
			Isolating switch	CNY/set
			Load switch	CNY/set
			Lightning arrester	CNY/piece
			Line erection	CNY/km
			Optical cable erection	CNY/km
			Transportation of line erection material	CNY/t
3		Earthing		
			Earthing bus	CNY/t or CNY/m
			Earthing electrode	CNY/piece
			Resistance-reducing agent	CNY/t
Ⅲ	Step-up substation equipment and installation			
1		Main transformer system		
			Main transformer	CNY/set
			Neutral equipment of main transformer (including isolating switch, current transformer and lightning arrester, etc.)	CNY/set
2		Distribution equipment		
			Circuit breaker	CNY/set
			Isolating switch	CNY/set
			Post insulator	CNY/piece
			Voltage transformer	CNY/set
			Current transformer	CNY/set
			Lightning arrester	CNY/set
			GIS	CNY/bay
			High voltage switchgear	CNY/set
			Arc suppression device	CNY/set
			Earthing transformer and earthing resistor	CNY/set

Table A.2 *(continued)*

No.	Level 1	Level 2	Level 3	Techno-economic indicator
			Bus	CNY/m or CNY/span
3		Reactive power compensation system		
			Reactive power compensator	CNY/set
4		Station service (emergency) power supply system		
			Station transformer	CNY/set
			Low voltage switchgear	CNY/set
			Distribution box	CNY/set
			Diesel generator	CNY/set
5		Power cables		
			Power cable	CNY/m
			Terminal	CNY/set
			Cable tray support	CNY/t or CNY/m
			Embedded pipe	CNY/m
			Fire protection	CNY/t or CNY/m or CNY/m^2
			Wall bushing	CNY/piece
6		Earthing		
			Earthing bus	CNY/m
			Earthing electrode	CNY/piece
			Resistance-reducing agent	CNY/t
7		Monitoring system		
			Substation monitoring system	CNY/set
			Relay protection and safety automatic device system	CNY/set
			Image monitoring system	CNY/set
			Fire alarm system	CNY/set
			Disturbance recorder system	CNY/set

Table A.2 *(continued)*

No.	Level 1	Level 2	Level 3	Techno-economic indicator
			Control cable	CNY/km
			Automatic power control system	CNY/set
			Wind power prediction system	CNY/lot
8		AC/DC system		
			UPS	CNY/set
			AC switchboard	CNY/set
			Charge (floating charge) device	CNY/set
			Discharge device	CNY/set
			DC panel	CNY/set
			Storage battery	CNY/set
9		Communication system		
			Communication equipment	CNY/set
			Communication power supply	CNY/set
			Communication cable / optical cable	CNY/km
10		Remote control and metering system		
			Remote control system	CNY/lot
			Metering system	CNY/lot
11		Subsystem testing		
			Transformer system testing	CNY/system
			Station power supply system testing	CNY/station
			Substation DC power supply system testing	CNY/station
			Bus system testing	CNY/segment
			Substation central signal system testing	CNY/station
			Fault filter system testing	CNY/station

Table A.2 *(continued)*

No.	Level 1	Level 2	Level 3	Techno-economic indicator
			Substation emergency lighting and UPS system testing	CNY/station
			Substation monitoring and five-prevention system testing	CNY/station
12		Electric system commissioning		
			Substation electric system testing	CNY/station
			Substation monitoring system testing	CNY/station
13		Special electrical test		
			Partial discharge voltage withstand test of transformer	CNY/set
			Transformer AC withstand voltage test	CNY/set
			Circuit breaker AC withstand voltage test	CNY/set
			GIS AC withstand voltage test	CNY/bay
			Power cable AC withstand voltage test	CNY/loop
Ⅳ	Other equipment and installation			
1		Heating, ventilation and air-conditioning system		
			Heating equipment	CNY/set
			Air-conditioning equipment	CNY/set
			Ventilation equipment	CNY/set
			Ventilation duct	CNY/m^2
2		Outdoor lighting system		CNY/lot
3		Firefighting, water supply and drainage system		
			Firefighting equipment	CNY/set
			Water supply and drainage equipment	CNY/set (group)

Table A.2 *(continued)*

No.	Level 1	Level 2	Level 3	Techno-economic indicator
			Pipeline	CNY/m
4		Occupational health and safety equipment		CNY/lot
5		Vehicles for O&M		CNY/lot
6		Amortization of equipment in control center		CNY/lot

Table A.3 shows the breakdown of civil works.

Table A.3 Breakdown of civil works

No.	Level 1	Level 2	Level 3	Techno-economic indicator
I	Wind farm			
1		Foundation works for wind turbine		
			Earth excavation	CNY/m^3
			Rock excavation	CNY/m^3
			Earth and rock backfill	CNY/m^3
			Masonry works	CNY/m^3
			Concrete	CNY/m^3
			Steel bar fabrication and installation	CNY/t
			Steel structure fabrication and installation	CNY/t
			Foundation pipe embedding	CNY/m
			Foundation grouting	CNY/m^3
			Pile	CNY/m or CNY/m^3
			Concrete antiseptic treatment	CNY/m^2
2		Wind turbine outgoing line works		
			Earth excavation	CNY/m^3
			Rock excavation	CNY/m^3
			Earth and rock backfill	CNY/m^3
			Sanding and brick covering	CNY/m

Table A.3 *(continued)*

No.	Level 1	Level 2	Level 3	Techno-economic indicator
			Pipe embedding	CNY/m
3		Foundation works for wind turbine transformer		
			Earth excavation	CNY/m^3
			Rock excavation	CNY/m^3
			Earth and rock backfill	CNY/m^3
			Masonry works	CNY/m^3
			Concrete	CNY/m^3
			Steel bar fabrication and installation	CNY/t
			Fencing	CNY/m
4		Earthing of wind turbine and transformer		
			Earth excavation	CNY/m^3
			Rock excavation	CNY/m^3
			Earth and rock backfill	CNY/m^3
			Earth replacement and backfill	CNY/m^3
Ⅱ	Collection line works			
1		Civil works of collection cable line		
			Earth excavation	CNY/m^3
			Rock excavation	CNY/m^3
			Earth and rock backfill	CNY/m^3
			Sanding and brick covering	CNY/m
			Pipe embedding	CNY/m
			Cable trench	CNY/m^3 or CNY/m
			Masonry works	CNY/m^3
			Concrete	CNY/m^3
			Steel bar fabrication and Installation	CNY/t
2		Civil works of overhead collection line		
			Earth excavation	CNY/m^3
			Rock excavation	CNY/m^3

Table A.3 *(continued)*

No.	Level 1	Level 2	Level 3	Techno-economic indicator
			Earth and rock backfill	CNY/m^3
			Masonry works	CNY/m^3
			Concrete	CNY/m^3
			Steel bar fabrication and installation	CNY/t
			Pile	CNY/m
3		Civil works of earthing for overhead collection line		
			Earth excavation	CNY/m^3
			Rock excavation	CNY/m^3
			Earth and rock backfill	CNY/m^3
			Earth replacement and backfill	CNY/m^3
Ⅲ	Step-up substation			
1		Site grading		
			General site grading	CNY/m^2
			Earth excavation	CNY/m^3
			Rock excavation	CNY/m^3
			Earth and rock backfill	CNY/m^3
			Masonry works	CNY/m^3
2		Foundation works of main transformer		
			Earth excavation	CNY/m^3
			Rock excavation	CNY/m^3
			Earth and rock backfill	CNY/m^3
			Concrete	CNY/m^3
			Steel bar fabrication and installation	CNY/t
3		Foundation works for reactive power compensator		
			Earth excavation	CNY/m^3
			Rock excavation	CNY/m^3
			Earth and rock backfill	CNY/m^3
			Concrete	CNY/m^3
			Steel bar fabrication and installation	CNY/t

Table A.3 *(continued)*

No.	Level 1	Level 2	Level 3	Techno-economic indicator
4		Foundation works for power distribution equipment		
			Earth excavation	CNY/m^3
			Rock excavation	CNY/m^3
			Earth and rock backfill	CNY/m^3
			Masonry works	CNY/m^3
			Concrete	CNY/m^3
			Steel bar fabrication and installation	CNY/t
5		Structure works for power distribution equipment		
			Concrete frame and bracket	CNY/m^3
			Fabrication and installation of steel frame and bracket	CNY/t
			Cable trench	CNY/m^3 or CNY/m
			Lightning arrester (tower)	CNY/t or CNY/set
			Emergency oil pond	CNY/base or CNY/m^3
6		Production buildings		
			Central control room (building)	CNY/m^2
			Distribution room (building)	CNY/m^2
			Reactive power compensator room	CNY/m^2
7		Auxiliary production buildings		
			Sewage treatment room	CNY/m^2
			Fire pump room	CNY/m^2
			Firefighting equipment room	CNY/m^2
			Diesel generator room	CNY/m^2
			Boiler room	CNY/m^2
			Storehouse	CNY/m^2
			Garage	CNY/m^2

Table A.3 (continued)

No.	Level 1	Level 2	Level 3	Techno-economic indicator
8		Site offices and living quarters		
			Office	CNY/m^2
			Duty room	CNY/m^2
			Dormitory	CNY/m^2
			Canteen	CNY/m^2
			Guard room	CNY/m^2
9		Outdoor works		
			Wall	CNY/m
			Gate	CNY/m^2 or CNY/Nos.
			Site road	CNY/m^2
			Site ground hardening	CNY/m^2
			Plantation	CNY/m^2
			Other outdoor works	%
Ⅳ	Road works			
1		Access roads		
			Earth excavation	CNY/m^3
			Rock excavation	CNY/m^3
			Earth and rock backfill	CNY/m^3
			Masonry	CNY/m^3
			Concrete	CNY/m^3
			Steel bar fabrication and installation	CNY/t
			Subgrade and cushion	CNY/m^2
			Surface	CNY/m^2
			Culvert	CNY/m
			Auxiliary facilities	CNY/m
2		Site roads		
			Earth excavation	CNY/m^3
			Rock excavation	CNY/m^3
			Earth and rock backfill	CNY/m^3
			Masonry	CNY/m^3
			Concrete	CNY/m^3

Table A.3 *(continued)*

No.	Level 1	Level 2	Level 3	Techno-economic indicator
			Steel bar fabrication and installation	CNY/t
			Subgrade and cushion	CNY/m^2
			Surface	CNY/m^2
			Culvert	CNY/m
			Auxiliary facilities	CNY/m
V	Other works			
1		Environmental protection		CNY/lot
2		Soil and water conservation		CNY/lot
3		Occupational health and safety		CNY/lot
4		Safety monitoring		CNY/lot
5		Water supply for firefighting, production and living		CNY/lot
6		Flood (tide) control		CNY/lot
7		Amortization of centralized production and operation management facilities		CNY/lot

Table A.4 shows the breakdown of other cost items.

Table A.4 Breakdown of other cost items

No.	Level 1	Level 2	Level 3	Techno-economic indicator
I	Land use cost			
1		Cost of land use		
			Land requisition fee	CNY/m^2
			Temporary land requisition fee	CNY/m^2
			Compensation for attachments on the land	CNY/lot
			Site clearing expense	CNY/lot
II	Preliminary work cost			CNY/lot
III	Project construction management cost			
1		Construction management cost		CNY/lot

Table A.4 *(continued)*

No.	Level 1	Level 2	Level 3	Techno-economic indicator
2		Construction supervision cost		CNY/lot
3		Consulting service fee		CNY/lot
4		Techno-economic review fee		CNY/lot
5		Project quality inspection and testing fee		CNY/lot
6		Norm preparation and management cost		CNY/lot
7		Project acceptance cost		CNY/lot
8		Construction insurance		CNY/lot
Ⅳ	Production preparation cost			
1		Costs of production staff training and advance mobilization		CNY/lot
2		Purchase cost of management tools, appliances and furniture		CNY/lot
3		Purchase cost of spare parts		CNY/lot
4		Cost of commissioning		CNY/lot
V	Scientific research, investigation and design fee			
1		Scientific research and test fee		CNY/lot
2		Investigation and design fee		
			Investigation fee	CNY/lot
			Design fee	CNY/lot
3		As-built drawing preparation cost		CNY/lot
Ⅵ	Other taxes and fees			
1		Compensation for soil and water conservation		CNY/lot

Annex B
(normative)
Documentation format of cost estimate

B.1 The format of the title page with signature and seal is shown in Figure B.1.

Approved by:

Verified by:

Reviewed by:

(Special seal of the registered cost engineer)

Checked by:

(Special seal of the registered cost engineer)

Prepared by:

(Special seal of the registered cost engineer)

Figure B.1 Format of the title page with signature and seal

B.2 The format for main techno-economic indicators is shown in Table B.1.

Table B.1 Main techno-economic indicators

<table>
<tr><td>Project name</td><td colspan="2"></td><td colspan="2">Price of wind turbine</td><td>CNY/kW</td><td></td></tr>
<tr><td>Location of construction site</td><td colspan="2"></td><td colspan="2">Price of tower</td><td>CNY/t</td><td></td></tr>
<tr><td>Designer</td><td colspan="2"></td><td colspan="2">Cost of wind turbine foundation</td><td>CNY/set</td><td></td></tr>
<tr><td>Owner</td><td colspan="2"></td><td colspan="2">Step-up substation</td><td>CNY/set</td><td></td></tr>
<tr><td>Installed capacity</td><td>MW</td><td></td><td rowspan="6">Quantity of main works</td><td>Earth and rock excavation</td><td>m^3</td><td></td></tr>
<tr><td>Unit capacity</td><td>kW</td><td></td><td>Earth and rock backfill</td><td>m^3</td><td></td></tr>
<tr><td>Annual on-grid electricity</td><td>kW · h</td><td></td><td>Concrete</td><td>m^3</td><td></td></tr>
<tr><td>Annual equivalent full-load hours</td><td>h</td><td></td><td>Steel bars</td><td>t</td><td></td></tr>
<tr><td>Static investment</td><td>CNY</td><td></td><td>Tower</td><td>t</td><td></td></tr>
<tr><td>Interest during construction</td><td>CNY</td><td></td><td>Pile</td><td>Piece or m or m^3</td><td></td></tr>
<tr><td>Total investment</td><td>CNY</td><td></td><td rowspan="2">Construction land area</td><td>Permanent land</td><td>ha</td><td></td></tr>
<tr><td>Static investment per kW</td><td>CNY/kW</td><td></td><td>Temporary (rental) land</td><td>ha</td><td></td></tr>
<tr><td>Investment per kW</td><td>CNY/kW</td><td></td><td colspan="2">Total construction period</td><td>month</td><td></td></tr>
<tr><td>Investment per kWh</td><td>CNY/ (kW · h)</td><td></td><td colspan="2">Manning quotas for production unit</td><td>person</td><td></td></tr>
</table>

B.3 The format for the table of cost estimate shall conform to that of Table B.2 to Table B.7.

Table B.2 Summary cost estimate

No.	Items	Equipment cost (CNY)	Costs of civil works and installation works (CNY)	Other cost items (CNY)	Total (CNY)	Percentage in total cost (%)
I	Auxiliary works					
1	…					
2	…					
II	Equipment and installation works					
1	…					
2	…					
III	Civil works					

Table B.2 *(continued)*

No.	Items	Equipment cost (CNY)	Costs of civil works and installation works (CNY)	Other cost items (CNY)	Total (CNY)	Percentage in total cost (%)
1	…					
2	…					
Ⅳ	Other cost items					
1	Land use cost					
2	Preliminary work cost					
3	Project construction management cost					
4	Production preparation cost					
5	Scientific research, investigation and design fee					
6	Other taxes and fees					
	Total cost of items in Ⅰ, Ⅱ, Ⅲ and Ⅳ					
Ⅴ	Basic contingency					
	Static investment Total cost of items in Ⅰ, Ⅱ, Ⅲ, Ⅳ and Ⅴ					
Ⅵ	Contingency for price variation					
Ⅶ	Interest during construction					
Ⅷ	Total investment of items in Ⅰ, Ⅱ, Ⅲ, Ⅳ, Ⅴ, Ⅵ, and Ⅶ					
	Static investment per kW (CNY/kW)					
	Investment per kW (CNY/kW)					

NOTE 1 Items down to Level 2 shall be listed in this table.
NOTE 2 Costs of civil works and installation works in this table include the installation material cost.

Table B.3 Cost estimate of auxiliary works

No.	Item	Unit	Quantity	Unit price (CNY)	Total (CNY)

NOTE Items down to Level 3 shall be listed in this table.

Table B.4 Cost estimate of equipment and installation works

No.	Name and specification	Unit	Quantity	Unit price (CNY)				Total (CNY)			
				Equipment cost	Installation cost			Equipment cost	Installation cost		
					Installation cost	Cost of installation material provided by the project owner	Subtotal		Installation cost	Cost of installation material provided by the project owner	Total

NOTE 1 Items down to Level 3 shall be listed in this table.
NOTE 2 Installation cost in this table includes the expenses for installation materials not provided by the owner.

Table B.5 Cost estimate of civil works

No.	Item	Unit	Quantity	Unit price (CNY)	Total (CNY)

NOTE Items down to Level 3 shall be listed in this table.

Table B.6 Estimate of other cost items

No.	Item	Unit	Quantity	Rate/Unit price (%/CNY)	Total (CNY)

Table B.7 Summary of annual cost

Unit: CNY

No.	Item	Total cost	Construction period			
			Year 1	Year 2	Year 3	…
Ⅰ	Auxiliary works					
Ⅱ	Equipment and installation					
Ⅲ	Civil works					
Ⅳ	Other cost items					
	Sum of Ⅰ, Ⅱ, Ⅲ, Ⅳ					
Ⅴ	Basic contingency					
Ⅵ	Static investment					
Ⅶ	Contingency for price variation					
Ⅷ	Interest during construction					
Ⅸ	Total investment					

B.4 The format of the attached tables shall conform to Table B.8 to Table B.13.

Table B.8 Summary of unit price of installation

Unit: CNY

No.	Item	Unit	Unit price	Direct cost					Indirect cost	Profit	Tax
				Basic direct cost			Installation material cost	Other direct cost			
				Labor cost	Material cost	Machinery cost					

Table B.9 Summary of unit price of civil works

Unit: CNY

No.	Item	Unit	Unit price	Direct cost				Indirect cost	Profit	Tax
				Basic direct cost			Other direct cost			
				Labor cost	Material cost	Machinery cost				

Table B. 10 Summary of machine-shift cost

Unit: CNY

No.	Name and specifications	machine-shift cost	In which					
			Depreciation cost	Repair cost	Assembling / disassembling cost	Labor cost	Power fuel cost	Other cost

Table B. 11 Unit price of raw materials of concrete

No.	Strength class of concrete	Cement grade	Gradation	Estimated quantity						Unit price (CNY)
				Cement (kg)	Admixture (kg)	Sand (m^3)	Crushed stone (m^3)	Additive (kg)	Water (kg)	

Table B. 12 Unit price of civil works

Code of norm: ________ ________ project Unit: ________

Method of construction:					
No.	**Item**	**Unit**	**Quantity**	**Unit price** (CNY)	**Total** (CNY)
I	Direct cost				
(I)	Basic direct cost				
1	Labor cost	man-day			
2	Material cost				
	…				
3	Machinery cost				
	…				
(II)	Other direct cost	%			
II	Indirect cost	%			
III	Profit	%			
IV	Tax	%			
V	Total	CNY			

Table B. 13 Unit price of installation works

Code of norm: ________ ________ project Unit: ________

Method of construction:					
No.	**Item**	**Unit**	**Quantity**	**Unit price** (CNY)	**Total** (CNY)
I	Direct cost				
(I)	Basic direct cost				
1	Labor cost	man-day			
2	Material cost				
	…				
3	Machinery cost				
	…				
4	Installation material cost				
(II)	Other direct costs	%			
II	Indirect cost	%			
III	Profit	%			
IV	Tax	%			
V	Total	CNY